Sasha The Kamikaze Cow Dog
℅ Permissions Coordinator
1215 Upper Powder River Rd.
Arvada, WY 82831

ISBN: 979-8-9854310-0-1

Library of Congress registration number: TX 8-947-396

Any reference to historical events, real people, or real places are used fictitiously. Any events described other than story base line and Sasha the dog, including names, characters, and places, are products of the author's imagination.

Illustrations by Andrea Murphy
Book authored and designed by Karlon Knudson

First Edition printed in 2019

Published by: Karlon Knudson
1215 Upper Powder River Rd.
Arvada, Wy 82831

www.sashathecowdog.square.site

*Dedicated to Kort, Wyatt,
Beau, and all the children of
the world who need a good
Sasha adventure in their lives!*

Sasha was a brand new puppy on the ranch, and she loved doing puppy things.

She loved playing Tug of War!

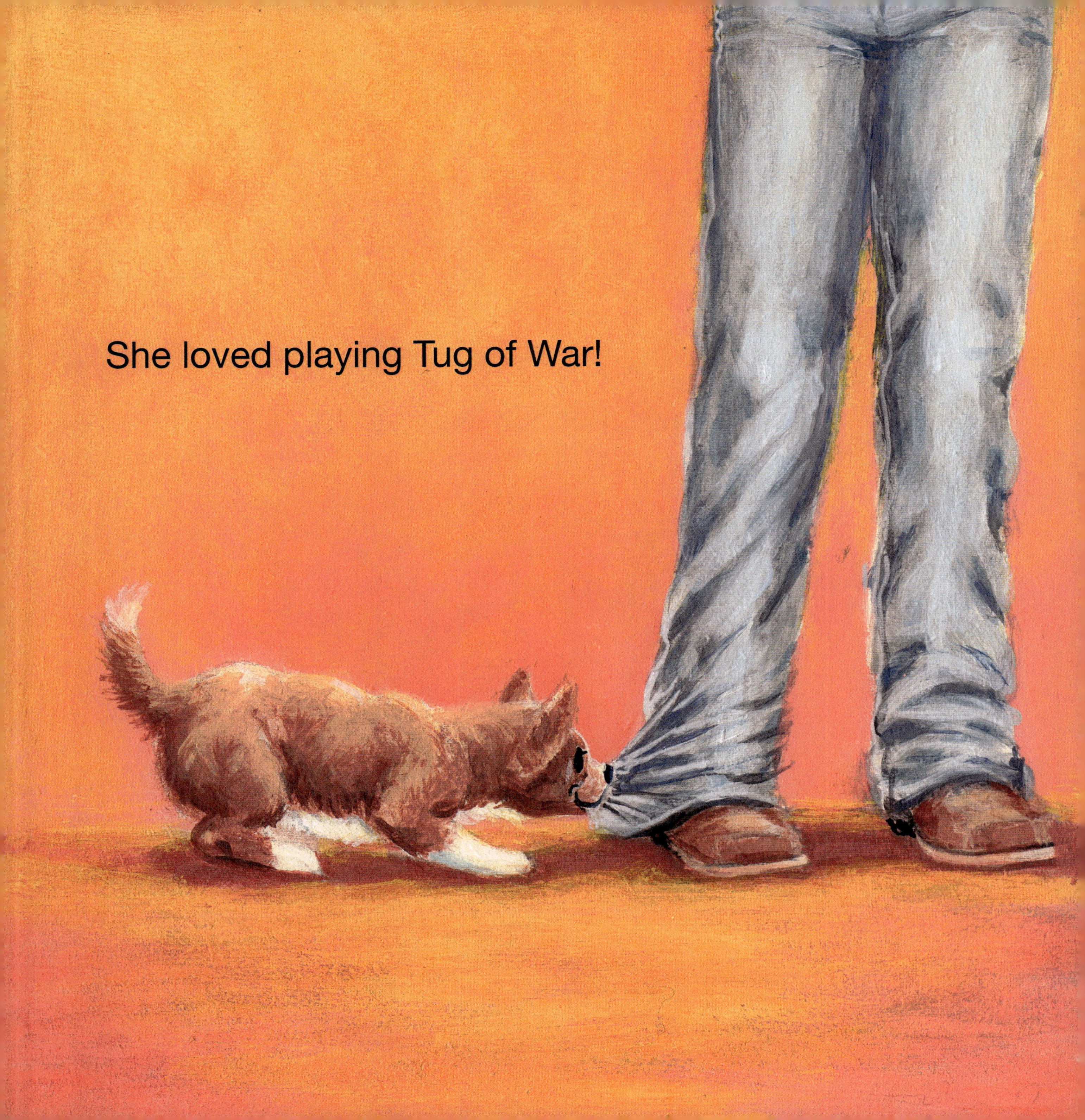

She chewed on squeaky toys,

and she wrestled with her friend, The Cowboy.

Every day she saw more things on the ranch. Everything was so new and exciting!

She had a great time running through the snow.

She got to ride the Four-Wheeler.

Her favorite thing to do on the ranch, was watching the other animals.

One day while feeding the animals with her friend, The Cowboy, she met the horses and the cats.

When she met the cows though, she couldn’t look away.

She thought, “It would be fun to swing on that tail!”

She wasn't afraid. She wanted to turn it into a game.......so that's exactly what she did.

Sasha thought, "What if the cows don't like my game?"

She decided to sneak up on one first, just to see if she could.

She got down really low behind the truck and very slowly crept towards one of the cows.

The cow had her head down eating hay and didn't see Sasha creeping up behind her.

When she was about five feet away, she couldn’t contain her excitement any longer.

She leapt for the cow's tail, grabbed it with her teeth, and held on!

The unsuspecting cow was so surprised by Sasha swinging from her tail, she spit her hay out, bellowed, and turned to face her attacker.

The startled cry from the cow and all the sudden motion, scared the other cows nearby. As Sasha let go of the cow's tail, she realized she was greatly outnumbered.

Sasha landed in front of the other cows, who were coming her way. The cow behind her was mad now, and also coming her way.

Sasha didn't know what to do! She ran left.....

She ran right……

She ran around a tree.......

She ran so fast, she ran right out of ideas!

Just as the cows were about to get her, Sasha squeezed under the back end of the truck and hid behind the tire.

As the cows ran by, Sasha thought,
"I don't think they like my game."

Sasha's friend, The Cowboy, laughed at all the commotion she had caused. As he reached down to rub her head he said, "You are kind of a kamikaze cow dog, aren't you!"

As she jumped back up on the truck to catch a ride home, Sasha couldn't help but wonder.....

"What might the other cows do
if I try to swing on their tails?"

About the Author

Karlon Knudson ranches in Wyoming. He draws his inspiration for writing from his son, who reminds him what it was like to be young. Sasha is also very inspirational as she brings daily adventures and laughs to their ranching lifestyle.

About the Illustrator

Andrea Murphy and her husband, Chance, live in Eastern Oregon. They are raising two smart and funny little cowboys named Wyatt and Beau. Most of her art depicts western lifestyles of Oregon, Montana, and Wyoming.

Made in the USA
Columbia, SC
30 October 2022

69708178R00020